How to Use Android Smartphones and Tablets
Creative Kiwis Tech Lab
Android Smartphones, Tablets

Published by Smashwords Inc

Copyright 2019 Bill Rosoman

All Rights Reserved,
No reproduction in any form without
the permission of the Author,
Bill Rosoman, communityhubhuntly@gmail.com

First Published 2019

ISBN   9781708487010

Table d Contents

Introduction to How to Use Android Smartphones and Tablets 2
My Current Android Device 2019 3
Tutorial One, Android Devices Introduction 5
Tutorial Two, Android Devices Intermediate 7
Installing Android Apps 7
Emergency Apps 8
Creating and Editing Documents in Android 9
Removing apps from your Device 10
Android House Keeping 10
Some Basic Shortcuts for most Programs 11
Tutorial Three, Android Devices Advanced 13
Other Tutorials 17
Other Online tutorials 23
Android Tutorials 23
Creative Kiwis Tech Lab 23
Computer/Technology/Devices Resources 24
Some Computer Rules 26
Certificate of Achievement Available 28
Final Thoughts 29
Bill Rosoman Biography 30
Assistance 32
Creative Kiwis, an Amazing Journey into the Future. 33
    Assistance 34

Introduction to How to Use Android Smartphones and Tablets

How to Use Android Smartphones and Tablet Devices

This is a multi media series of tutorials and information on learning to use Android Devices Basics. Whether it be a Smartphone or a Tablet or other Android device.

There is the Book, Ebook and many YouTube Videos to help

you get the most out of your Android Device. As well we offer online assistance.

These tutorials cover from the Basics of how to make a phone call or send a text message to using apps like Banking. Google maps, Screenshot, Hotspot, Stream/Cast, Google Photos etc.

Keywords
Android, Smartphone, Tutorial, Youtube, Android Smartphone Tutorial, Android, Smartphone, Use Android Apps, Ebook, Book, Learn to use an Android Smartphone, Learn to use an Android Tablet, Google Apps, Smartphone Basics.

Website www.creativekiwis.com/android-devices-tutorials

Email communityhubhuntly@gmail.com

Top of document

My Current Android Device 2019

Huawei Y7 Pro (2019)

Versions: DUB-LX2

NETWORK Technology
GSM / HSPA / LTE
LAUNCH Announced 2019, January
Status Available. Released 2019, January
BODY Dimensions 158.9 x 76.9 x 8.1 mm (6.26 x 3.03 x 0.32 in)
Weight 168 g (5.93 oz)
SIM Hybrid Dual SIM (Nano-SIM, dual stand-by)
DISPLAY Type IPS LCD capacitive touchscreen, 16M colors
Size   6.26 inches, 97.8 cm2 (~80.0% screen-to-body ratio)
Resolution 720 x 1520 pixels, 19:9 ratio (~269 ppi density)
PLATFORM OS Android 8.1 (Oreo); EMUI 8.2
Chipset Qualcomm SDM450 Snapdragon 450 (14 nm)
CPU Octa-core 1.8 GHz Cortex-A53
GPU Adreno 506
MEMORY Card slot microSD, up to 512 GB (uses shared SIM slot)
Internal 32 GB, 3 GB RAM
MAIN CAMERA Dual 13 MP, f/1.8, PDAF
2 MP, depth sensor
Features   LED flash, HDR, panorama
Video 1080p@30fps
SELFIE CAMERA Single 16 MP
Features   HDR
Video 1080p@30fps
SOUND   Loudspeaker Yes
3.5mm jack Yes
  Active noise cancellation with dedicated mic
COMMS WLAN   Wi-Fi 802.11 b/g/n, Wi-Fi Direct, hotspot
Bluetooth 4.2, A2DP, LE
GPS Yes, with A-GPS, GLONASS, BDS

NFC Yes (market dependent)
Radio FM radio
USB microUSB 2.0
FEATURES Sensors Accelerometer, proximity, compass
BATTERY Non-removable Li-Ion 4000 mAh battery
Charging Fast battery charging 5V/2A 10W
MISC Colors Black, Aurora
Models DUB-LX2

Cost $NZ350

Preview huawei y7 pro www.tinyurl.com/ckpreview

I only have a smartphone and a Laptop. Though I have access to other technology at the Local Community House were I am the IT Guy and the Computer Club.

I find I can do 90% of tasks I want to do on my phone.

Tutorial One, Android Devices Introduction

This is a Hands on Tutorial Please Bring your Smartphone and Tablet or Ipad.

There will be a Demonstration then you will be able to use your Device to do the same.

At the end of the Tutorial participants will receive a Certificate of Achievement

Introduction

Android Devices Basics www.tinyurl.com/ckbasics

The basics of an Android Smartphone

Overview of Huawei Y7 pro www.tinyurl.com/ckpreview
An overview of My Current Smartphone

Make Receive Phone Call www.tinyurl.com/ckphonecall

Make Receive Text SMS Message www.tinyurl.com/cktext

Make Receive Email www.tinyurl.com/cksendemail

Use Google Contacts www.tinyurl.com/ckcontacts

Use Google Chrome Web Browser

www.tinyurl.com/ckchrome

Chrome is used to cruise the Internet, find a recipe, find information.

Using the Camera www.tinyurl.com/ckcamera

Promo Video www.tinyurl.com/ckpromo

Other Online tutorials

https://edu.gcfglobal.org/en/subjects/tech/

https://edu.gcfglobal.org/en/topics/smartphonesandtablets/

https://edu.gcfglobal.org/en/androidbasics/

Upon completing the above Tutorial and you are are happy with your new skill levels please email your details to communityhubhuntly@gmail.com and we will arrange your certificate to be forwarded to you.

Tutorial Two, Android Devices Intermediate

This is a Hands on Tutorial Please Bring your Smartphone and Tablet or Ipad.

There will be a Demonstration then you will be able to use your Device to do the same.

At the end of the Tutorial participants will receive a Certificate of Achievement

Installing Android Apps

The easiest way is to use the App Store (Google Play) by Clicking on App Store Icon, Search for the App You want and Download and Install.

You can also download an .apk app from the internet and install just need to be careful it is not dangerous for your device.

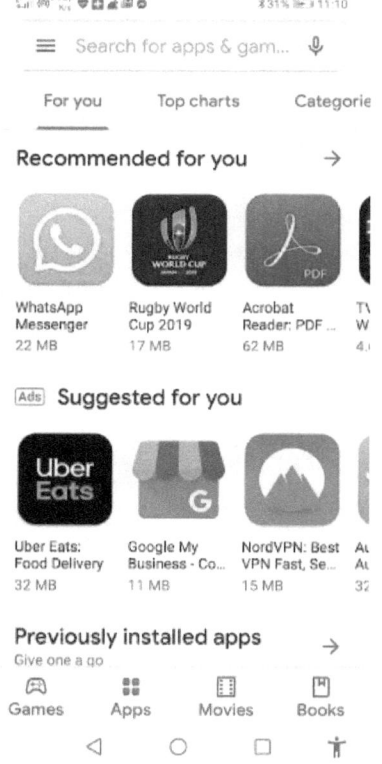

How to Install an App using Google Play Store

www.tinyurl.com/ckinstallapp

Emergency Apps

Emergencies www.tinyurl.com/ckemergency2

ICE Lock In Case of Emergency your personal details, medical details and next of kin
I'm Here   Panic Alarm (Sends a message with your location)
GPS Tools Various GPS Tools, compass, altimeter
Torch   For Night time use
Camera To record an accident or send pic to friends or Facebook
Weather Keep track of the weather
Red Cross First Aid First Aid Book
Red Cross Hazards Be warned about natural disasters, earthquakes land slides etc
Simple GPS Get your GPS Coordinates
Waze Maps Get Map Directions works offline
Promo Add A Promo from Creative Kiwis Tech Lab www.tinyurl.com/ckpromo

Most of the apps work offline and I have a solar panel in my van and a solar power bank to keep my phone charged.

Creating and Editing Documents in Android

Editing Documents is a bit harder in Android there are some limitations.

You can edit a text file and then email to yourself as an attachment, or

Just create an email, save as a draft and then email yourself when online.

I am using Jota Text Editor, it is very nice t take quick notes and being able to find them when needed.

You can use Google Drive or something like by using your Google Docs account online (cloud computing). Or WPS Office app.

www.tinyurl.com/ckwpsoffice WPS is Free and a great Office Suite very similar to Microsoft Office. Creates and uses PDF files as well.

https://docs.google.com/document/u/0/ Google Docs, you can use via a browser like Google Chrome or install the Google Docs Apps and use that way. Google Docs is Free.

Removing apps from your Device

Follow these steps to remove an application from the Tablet

Click on the "Settings" icon
Click on "Applications"
Click on "Manage Applications"
Click on the application you want to remove
Click on "Unintstall"
Click on "OK"
Wait for the application to uninstall
Click on "OK"
The application should now be uninstalled

www.tinyurl.com/ckremove

Android House Keeping

Download the Clean Master app from the Google Play Store

play.google.com/store?hl=en_US

Run Clean Master once a week to keep your device clean of all the rubbish you collect.

House Keeping www.tinyurl.com/ckcleanmaster

Clean Master - Antivirus, Applock & Phone Cleaner is a super antivirus cleaner app for Android with over 1 billion users. We provide powerful functions to speed up your phone, clean junk files, clean virus, protect your privacy & phone security.

Some Basic Shortcuts for most Programs

On an Android Device to Copy Text Hold Finger Down on Text, Select Text or Select All, Select Copy or Cut (Move to), Go to new Location Paste or Hold Down Finger and Paste. Ipad/Iphone Use Apple Control and Similar Commands.

How to Take Screen Shot www.tinyurl.com/ckscreen

Android Device as a hotspot www.tinyurl.com/ckhotspot

Using a hotspot allows your other devices to access the Internet from one device. I use a hotspot so my laptop can access the Internet.

Use a Wordpress Blog www.tinyurl.com/ckwordpress

https://asiancruising.wordpress.com/ is one of my blogs I use for traveling. Very easy to use on your phone or tablet and to upload photos or share videos.

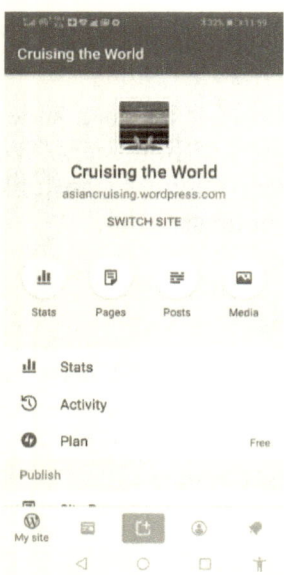

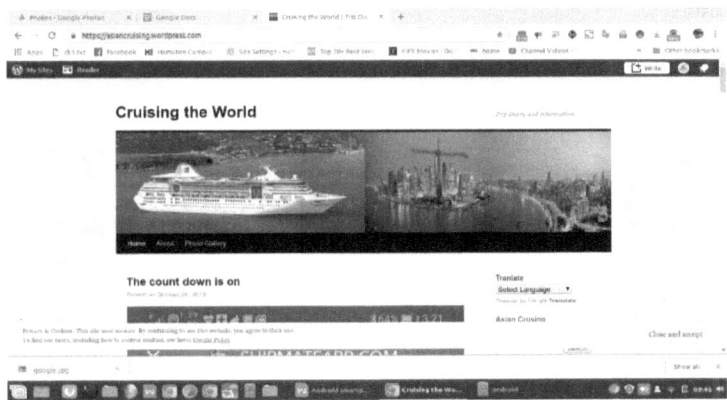

Using Google Photos www.tinyurl.com/ckgphotos

If your device is setup properly using Google Photos your photos are automatically uploaded to Google Photos in the Cloud.

https://photos.google.com/u/1/

So if you loose your device or it gets stolen you do not loose all your photos.

Having your photos online also makes it easy to show friends and family no matter were you are as long as you have access to the internet.

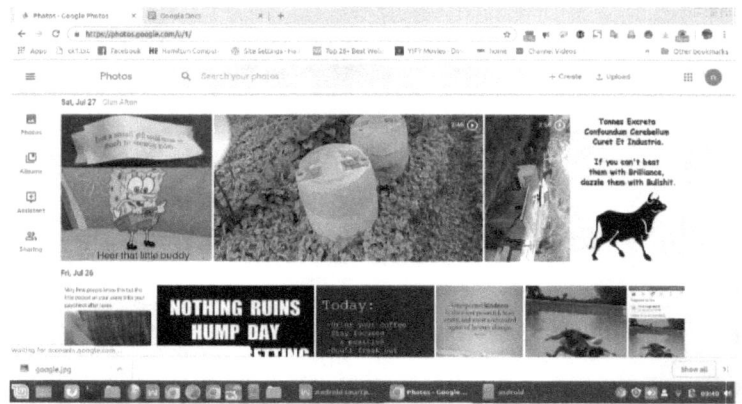

Upon completing the above Tutorial and you are are happy with your new skill levels please email your details to communityhubhuntly@gmail.com and we will arrange your certificate to be forwarded to you.

TOP

Tutorial Three, Android Devices Advanced

This is a Hands on Tutorial Please Bring your Smartphone and Tablet or Ipad.

There will be a Demonstration then you will be able to use your Device to do the same.

At the end of the Tutorial participants will receive a Certificate of Achievement

Find My Device www.tinyurl.com/ckfindmydevice

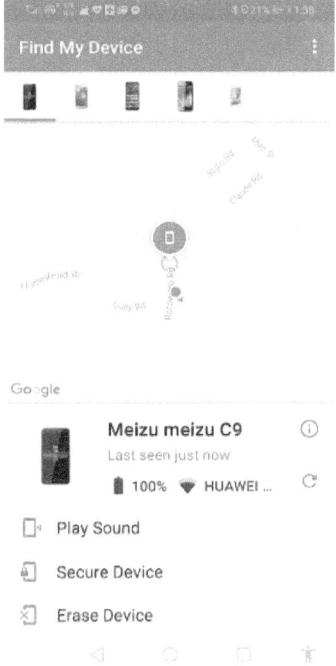

handy app to find phone if lost or stolen.

Android Device as Home Security
www.tinyurl.com/cksecurity

Security Camera in My Van, it Runs 24/7

 At Home Security App

Android Device Scanning Document or Picture app
www.tinyurl.com/ckscanning

Connect to WiFi www.tinyurl.com/ckconnect
How to connect your device to WiFi.

Streaming and Casting from Android Device to a TV
www.tinyurl.com/ck1stream

Android Devices Using Clock, Alarm, Time, Stopwatch
www.tinyurl.com/ckclock

Android Top 10 apps www.tinyurl.com/cktop10

Google Apps and Services www.tinyurl.com/ckgoogleapps

Most people do not realise that once you get a gmail account Google offers many other services for free.

https://mail.google.com/mail/u/0/#inbox

www.youtube.com

https://photos.google.com/u/1/

https://docs.google.com/document/u/0/

https://drive.google.com/drive/u/0/my-drive

https://contacts.google.com/u/0

www.google.om

https://calendar.google.com/calendar/r?tab=rc

https://translate.google.co.nz

https://earth.google.com/static/9.3.97.2/app_min.html

These are just a few of the Google Services.

Most are apps or you can use through Google Chrome Web Browser.

Upon completing the above Tutorial and you are are happy with your new skill levels please email your details to communityhubhuntly@gmail.com and we will arrange your certificate to be forwarded to you.

Other Tutorials

OK Google or Google assistant
Very handy for hands free or to ask Google a quick question etc.
www.tinyurl.com/ckokgoogle

Setup a Data Pack   www.tinyurl.com/cksetupdata

You need credit on your device to make phone calls and send SMS texts.

But to use the Internet you need a data pack.

BTW if you do not have a data pack your device balance will disappear very quick as you will be on casual rates.

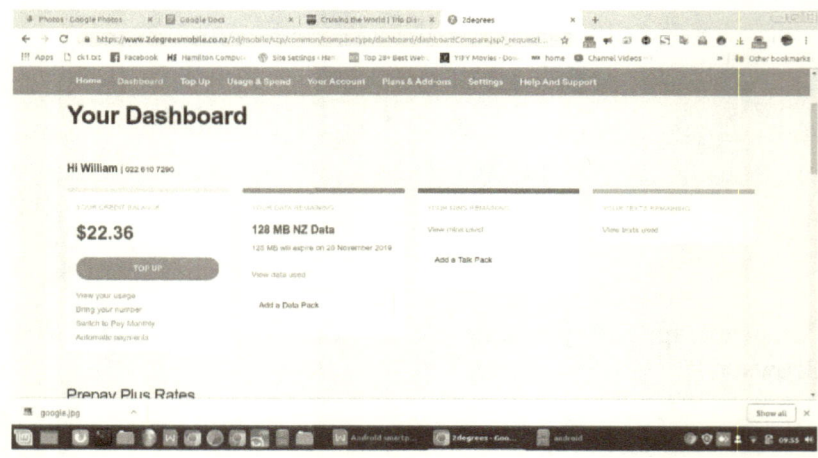

You can use your ISP provider app or use their website via a browser.

You need to setup a data package and usually you have to top up each month.

On my main phone I pay $70 to get 25gb of data a month this also gives me free phone calls and free texts. My data also carry's over from month to month if I do not use the data.

I have another few phones as Security cameras I get 500mb of data for $10 a month.

Using Google Calendar www.tinyurl.com/ckcalendar2

Find an Easter Egg www.tinyurl.com/ckeasteregg

Change Display Timeout Setting www.tinyurl.com/ckdisp_ay

Check Whether you are on WiFi or Data www.tinyurl.com/ckaccess

Using Google Earth www.tinyurl.com/ckgearth

Preview huawei y7 pro https://tinyurl.com/ckpreview

Using Google apps www.tinyurl.com/ck1gapps

Android Stream and Download Music www.tinyurl.com/ck1music

Streaming and casting from Android device to a TV www.tinyurl.com/ck1stream

my camper van www.youtube.com/watch?v=vaFeZMwxMz0

Install a Widget www.tinyurl.com/ckwidget

Use Jota Text Editor www.tinyurl.com/ckjota2

Use File Manager www.tinyurl.com/ckfileman

CK Promo https://tinyurl.com/ckpromo

Using Online Banking www.tinyurl.com/ckbanking

Enable Developer mode www.tinyurl.com/ckdevelop

Using WPS Office for Documents, Presentations, Spreadsheets, PDF
www.tinyurl.com/ckwpsoffice

Using GPS Tools   www.tinyurl.com/ckgpstools

   one of many GPS Tools

Download music from you tube www.tinyurl.com/ckytmusic

VLC music videos www.tinyurl.com/ckmedia2

Upload video to YouTube www.tinyurl.com/ckyoutubeup

Using Google maps or Waze Maps www.tinyurl.com/ckmapapps

Android use Anydesk www.tinyurl.com/ckanydesk

Android List note, Speech to Text App www.tinyurl.com/cklistnote

Upload Video to YouTube www.tinyurl.com/ckytupload

Android Internet Provider www.tinyurl.com/ckprovider

Android Devices News and TV apps www.tinyurl.com/cknewstv

Android Devices Travel apps www.tinyurl.com/cktravel

Android Devices Shopping apps www.tinyurl.com/ckshopapps

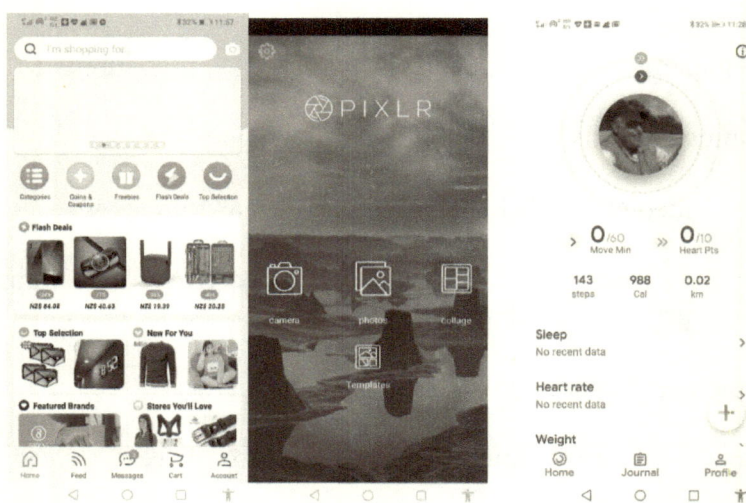

Aliexpress Shopping   Pixlr Picture Edit   Google Fit

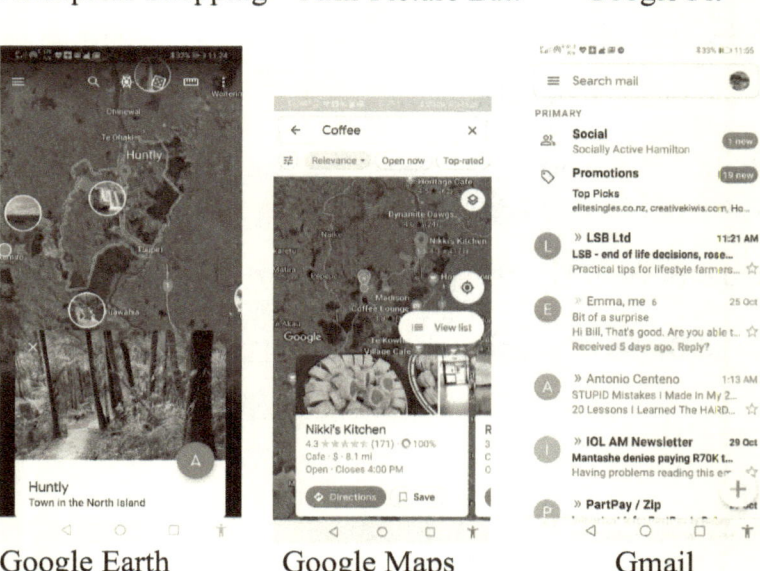

Google Earth   Google Maps   Gmail

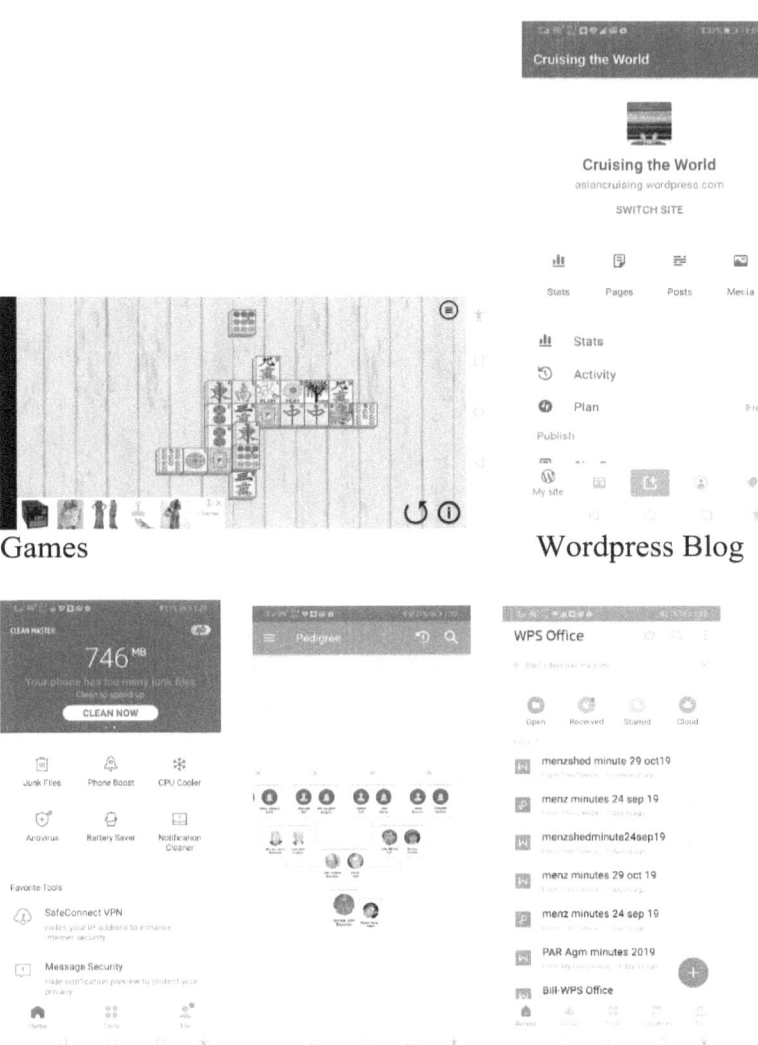

Games   Wordpress Blog

Clean Master (clear junk files) Family Tree WPS Office

Other Online tutorials

https://edu.gcfglobal.org/en/subjects/tech/

https://edu.gcfglobal.org/en/topics/smartphonesandtablets/

https://edu.gcfglobal.org/en/androidbasics/

Android Tutorials

www.tinyurl.com/ck1playlist

Creative Kiwis Tech Lab

Android Devices (like Tablets and Smartphones) Basic Tutorials and Videos

Online Tutorials and Assistance with Technology

Android Devices
Smartphones and Tablets

Facilitator Bill Rosoman Dip CS

Also available on a USB Drive and an ebook

www.creativekiwis.com

Apps I have ATM

2 Degrees	Asb	Trademe	ABC News
Adsense	Amazon	Aliexpress	Wish
G analytics	At home	BBC	Campermate
Cam scan	Ccleaner	Creative kiwis	Daily roads
dolphin browser	Facebook	Family search	Find my device
Firefox	Google fit	Flash fox	Fly buys
gallery	Gaspy	GPS tools	Group share
I'm here	Jota	Kiwibank	List note
Meetup	Messenger	New world	Open signal
Opera Browser	Paypal	Pb tech	Photoshop
Pixlr	Pinterest	Qr scan	Rankers
Radio NZ	Sky news	UK TV	Sound meter
Showbox	Talking clock	Tea TV	Warehouse
Threenow	Torque	Tube mate	TVNZ
VLC	WPS Office	Waze	Wordpress

And more I have 170 apps ATM.

TOP

Computer/Technology/Devices Resources

YouTube Channel   www.youtube.com/user/leftfieldnz

Android   www.tinyurl.com/androidvideos

Windows   www.tinyurl.com/windowsyoutube

Linux...   www.tinyurl.com/linuxyoutube

Cruise Around Australia   www.tinyurl.com/cruiseaus

Bills Tutorials   www.tinyurl.com/billtutorials

Computer Basics   www.tinyurl.com/thebasics2

Office Tutorials

www.gcflearnfree.org

www.tinyurl.com/officeonlinehelp

Free Software www.tinyurl.com/freesoft5

http://itsfoss.com/websites-for-foss

Google Free Stuff   www.tinyurl.com/googlefreestuff

Google Apps and Services   www.tinyurl.com/googleapps6

Computer Hub Huntly
http://computerhubhuntly.wixsite.com/home

Creative Kiwis an Amazing Journey
www.creativekiwis.com

Connect and Stream Smart TV www.tinyurl.com/smarttvvideo

Google Drive (Cloud Storage) http://drive.google.com/drive

How to avoid scams www.tinyurl.com/avoidscam6

Scam Advice and Reporting www.tinyurl.com/scamadvice

TOP

Some Computer Rules

Rule 1 THOU SHALL NOT USE (NEVER EVER);

MS Internet Explorer (Edge), MS Outlook, MS Outlook Express

Reason Too Many Bugs and made by Microsoft.

Rule 2 THOU SHALL USE (ALWAYS AND ALWAYS);

Use Google Chrome Internet Browsing

http://support.google.com/chrome/bin/answer.py?hl=en&answer=95346

Or Mozilla Firefox Internet Browsing http://download.mozilla.org/?lang=en-GB&product=firefox-19.0&os=win

Mozilla Thunderbird Email
https://www.mozilla.org/EN/thunderbird

Explorer++ to replace Windows File Explorer (a shocking MS program) http://explorerplusplus.com

Rule 3 THOU SHALL (AT LEAST WEEKLY/MONTHLY);

Run Spyware and Run Cclean As well as keep your virus checker up to date.

Rule 4 THOU SHALL (ALWAYS AND ALWAYS);

Use a gmail, yahoo or MS live email and not a NZ ISP email account.

If you change ISP or move overseas you loose all your emails.

Rule 5 THOU SHALL (ALWAYS AND ALWAYS);

Use Google Search when you have a problem.

Take a Deep Breath and have a Cup of Tea!!!!!! Ask a techno friend or Email Bill communityhubhuntly@gmail.com

Rule 6 THOU SHALT SAVE YOUR WORK OFTEN.

Please save often so you do not loose all your hard work

Rule 7 Thou Shalt backup data to the cloud

https://drive.google.com/drive/u/0/my-drive

The number of times I have heard people losing many hours work because they did not save their work is amazing.

Please save your work often. I also save different versions of documents doc1 doc2 etc. This prevents stuff ups that may occur.

Rule 8 THOU SHALL (ALWAYS AND ALWAYS);

HAVE FUN AND GIVE IT A GO, LOL! Kia Kaha

Rule 9 Thous Shalt not Forget Your Passwords.

Rule 10 Try to learn something new every day, with Technology do not get stuck in 1st gear but expand your horizons.

Rule 11 All Software is Free there is no need to buy and software, there is always a good free alternative.

Certificate of Achievement Available

A certificate is available once you have completed our tutorials.

Certificate of Achievement

this certificate is awarded To:

_____

For Completing a Creative Kiwis Tech Lab

Android Device Educational Course

_____   _____
Date   Signature

www.creativekiwis.com

Final Thoughts

Tablets will be really popular in coming years because of there low cost, portability, flexibility and great fun.

While I have a Laptop PC at the moment I use my smartphone to do about 90% of my computer stuff.

My main interest in an Android Device is;

Low Price
Portability
Flexibility
Fun
New Frontier
Low Power use (I have a Motor Home)
For travel

For Ebooks
For Movies
For Music

Be careful and have some fun.

Great for the kids too!

Bill Rosoman Biography

William John Rosoman Dip C.S
New Zealand. Aotearoa.
Telephone : +64-21-233-5427
Email communityhubhuntly@gmail.com

www.creativekiwis.com

Bills Skills:
Bill has excellent DIY Skills. Bill built his own two bedroom home in 1979. In 2009 and again in 2019 converted a Truck to a Mobile Home,

Bill has been using computers since the early 1980s. Bill is a Self-Motivated Learner and accept tasks as a Personal Challenge. Bill has had Email and Internet type access to the world since the mid 1980s.

Bill has 30+ years experience using Microsoft Products like; Windows 95/98/ME/XP/Vista, Office, Word, Excel, Powerpoint, Publisher as well as MYOB Accounting, Web Design, Graphics, CD/DVD Tools, Linux etc.

Bill has been using Linux Mint OS in preference to Windows

OS for many years now.

Bill has been involved in Business Administration, Marketing and Planning as well as Retail for over 30 years.

Bill was self-employed working with small businesses (Private Training Establishments) doing day to day Office Work and Preparing Weekly/Monthly/Yearly Wages/Accounting Information for Management to NZQA/IRD Audit Standard. As well as writing and preparing documentation for NZQA.

Bill has many years retail experience. Working in a Tavern, Craft Shop, Café, Fish and Chip shop.

Bill has a Diploma in Computing, Stream Support, Level 5.

Bill is ambidextrous and also a left and right brain thinker.

Bill has Drivers License is for Classes 1, F, R, W.

Bill has the OSH Certificate for my Forklift License.

Bill has the St John Basic First Aid Certificate.

Bill has Desktop Publishing Experience. Using Free Open Source Software (FOSS) Scribus, Open Office, WPS Office, Darktable, Calibre, Gimp and Inkscape.

Bill has a passion for computing, telecommunications and the internet as well as electronics and things electrical.
Bill is a successful Book/Ebook Publisher on www.smashwords.com/profile/view/leftfieldnz
Bill is a Vblogger on YouTube www.youtube.com/leftfieldnz
Bill has been specialising in doing book layout, book covers

and getting the books and ebooks published online and selling

If Bill can help you get your Book or Ebook Online or if you need a Website or similar contact Bill now
communityhubhuntly@gmail.com

Assistance

If you Need Assistance try a Google Search

www.google.com

Or Contact;

communityhubhuntly@gmail.com
computerhubhuntly.wixsite.com/home
Phone +64 28 255 36107
Kia Kaha Bill

My Place

Creative Kiwis, an Amazing Journey into the Future.

Books, Ebooks, Audio Books and much more
www.creativekiwis.com

Bill Rosoman ebooks on Smashwords
http://www.smashwords.com/profile/view/leftfieldnz

Craig Lock ebooks on Smashwords
https://www.smashwords.com/profile/view/craiglock

Creative Kiwis Videos at
www.youtube.com/leftfieldnz

Creative Kiwis Blog
http://leftfieldnz.wordpress.com/

Assistance

If you Need Assistance Contact;
communityhubhuntly@gmail.com
computerhubhuntly.wixsite.com/home
Phone 028 255 36107
Kia Kaha Bill

Copies of this ebook from Smashwords:
www.smashwords.com/profile/view/leftfieldnz
Website: www.creativekiwis.com
Facebook: www.facebook.com/leftfieldnz
Blog: http://leftfieldnz.wordpress.com

An ebook,

Keywords: Creative Kiwis, an Amazing Journey, Creative, Kiwis, New Zealand,
Android, Android Devices, Smartphone, Tablet, Tutorial, Training

#####

www.ingramcontent.com/pod-product-compliance
Lightning Source LLC
Chambersburg PA
CBHW030547220526
45463CB00007B/3019